Reading Es

LIVING WONDERS

STAYING ALIVE
Regulation and Behavior

SUSAN GLASS

PERFECTION LEARNING®

Editorial Director: Susan C. Thies

Editor: Mary L. Bush

Design Director: Randy Messer

Book Design: Michelle Glass
Lori Gould

Cover Design: Michael A. Aspengren

A special thanks to the following for his scientific review of the book:

Paul Pistek, Instructor of Biological Sciences, North Iowa Area Community College

Image Credits:

Tom Brakefield/CORBIS: p. 24 (middle); Martin Harvey/CORBIS: p. 25 (top); Anthony Bannister/Gallo Images/CORBIS: p. 26 (top); Paul A. Souders/CORBIS: p. 34 (bottom); Richard Hamilton Smith/CORBIS: p. 37 (top); Clive Druett/Papilio/CORBIS: p. 39 (bottom); Lynda Richardson/CORBIS: p. 40 (top); Jonathan Blair/CORBIS: p. 40 (bottom)

Photos.com: front cover, back cover, all Try This! backgrounds, background on all pages, pp. 2, 4 (top & middle), 5 (top left & right), 6, 7 (bottom), 8, 9, 10, 11 (bottom), 12 (top), 13, 14 (bottom), 15, 16 (top & bottom), 17, 18, 19 (top & middle), 20, 21, 22, 23 (bottom), 24 (top & bottom), 25 (left), 26 (right), 27, 28, 29, 30 (bottom), 31, 32 (top & bottom), 33, 34 (top), 35, 36, 37 (middle), 38, 39 (top), 41, 42, 44, 46, 47, 48; Digital Stock Photography: pp. 3 (left & right), 12 (left), 14 (top), 16 (middle), 19 (bottom), 30 (top), 32 (middle); Corbis Royalty Free: p. 4 (bottom); ClipArt.com: p. 5 (bottom); Corel: pp. 7 (left), 23 (top); Perfection Learning Corporation: pp. 11 (top), 37 (left)

For information, contact
Perfection Learning® Corporation
1000 North Second Avenue, P.O. Box 500
Logan, Iowa 51546-0500.
Phone: 1-800-831-4190
Fax: 1-800-543-2745
perfectionlearning.com

2 3 4 5 6 7 BA 10 09 08 07 06 05
Paperback ISBN 0-7891-6325-x
Reinforced Library Binding ISBN 0-7569-4482-1

CONTENTS

chapter one

Staying Alive

LIFE IS HARD

• • •

IT'S A TOUGH WORLD OUT there. Mother Nature never lets things stay the same for long. Seasons change. Volcanoes erupt. Earthquakes shake the ground. Tornadoes blow across the land. For a living thing to survive, it has to adjust to constant changes in its surroundings.

The goal of all **organisms** is to stay alive. They have to find food, protect themselves and their **offspring**, and **reproduce** so their **species** can continue. And they must do all this in a constantly changing **environment**. What is an organism's environment? An environment is all of the factors in a plant's or animal's surroundings. This includes food and water sources, light, shelter, weather, and all other living things.

But organisms have to do more than just adapt to their outside environment. They also have to maintain the right conditions inside their bodies. Living things have to sense changes inside and outside their bodies and react to what's going on. This is known as **regulation**. The way an organism responds to its internal (inside) and external (outside) environment is its **behavior**. A hot elephant spraying itself with water, a squirrel hiding acorns, and a person pulling a hand away from a flame are all examples of behaviors. These actions help the organisms survive in their environment.

WHO LIVES AND WHO DIES?

Staying alive is a lot of work. A lot of species haven't made it. When they couldn't keep up with the changing environment or maintain their own body conditions, they became **extinct**.

Charles Darwin was an English scientist who formed a theory about the survival of species. He called his ideas "natural selection." Darwin believed that the organisms with characteristics, or traits, that help them survive in their environment would continue to reproduce and carry on their species. Those with less helpful traits would eventually die off. In other words, certain organisms are selected by nature to pass on their traits to future generations that will maintain the species.

Sloth

An organism's traits can include many things. Color is a trait. Shape is a trait. A deadly bite or poisonous leaves are traits. Every organism has its own unique set of traits. An organism's regulation and behaviors are an important part of that set. These traits help determine who or what lives and dies in the game of life.

Survival of the Fittest

Darwin's theory of evolution by natural selection is also known as "survival of the fittest." In biology (life science), "fitness" refers to an organism's ability to survive and reproduce. Darwin's "fittest" aren't necessarily the biggest, fastest, or strongest. Instead they may be the small mice that burrow underground to keep from freezing in the winter, the sloths that move so slowly that no **predators** notice them, or the tiny plants that wind their way up trees so they can reach the sunlight. These organisms will survive because they can regulate and behave in ways that enable them to survive. They are best fit for their environment.

Green mamba snakes have very poisonous venom that can kill victims quickly.

chapter two

Getting It Just Right

Dogs are warm-blooded mammals.

REMEMBER GOLDILOCKS? THE porridge she ate at the three bears' house couldn't be too hot or too cold. It had to be "just right." The same is true of organisms. Their internal conditions can't be too hot or too cold. They must be "just right" in order for them to survive. This is called *homeostasis*.

Regulation is the ability to maintain healthy conditions inside a body even when conditions outside the body change. Every organism requires different internal conditions. The average temperature for a human being is different than that of a horse or a sparrow or a plant. Warm-blooded animals maintain a constant body temperature regardless of the air temperature. Cold-blooded animals have body temperatures that change with the air temperature. When it's cold outside, their body temperature decreases. When air temperature increases, so does their body temperature. However, cold-blooded animals must still stay within a range of temperatures to be healthy.

Tortoises are cold-blooded reptiles.

What Is the Average Body Temperature of a Horse?

Most warm-blooded **mammals** have approximately the same average body temperature as humans—98.6°F. There are some exceptions though. A few are shown below.

Goats. . . . . 103.1°F
Rabbits. . . . . 102°F
Cows . . . . 101.3°F
Cats and Dogs. . . . . 101°F
Horses . . . . 100.4°F
Whales . . . . 95.9°F

**Temperatures are approximate and will vary within a given range of a degree or two.

Since most environments have varying weather conditions, organisms must have strategies or behaviors for responding to these changes. Cold winters become steamy summers. A nap under a shady tree might be followed by a walk on a sunny beach. Even the switch from cool nights to warm days can be tricky for some organisms. But successful organisms sense changes inside and outside the body and react to them. When they get too hot, they find ways to cool down. When they get too cold, they find ways to warm up. That's what regulation is all about—getting and staying "just right."

When temperatures rise above 59°F, walruses stay in the water to keep cool.

REGULATION IN HUMANS

• • •

Because humans are warm-blooded, their internal body temperature has to stay the same through summer heat waves and winter snowstorms. Luckily, the human body has several ways of cooling off or warming up.

What happens if you have to walk to school on a freezing-cold morning and you forget your hat and gloves? Halfway down the block, the cold sensors in your skin send messages to your brain like "Danger! Your body temperature is going to drop too low if you don't do something now!" The brain responds and sends messages to the skin telling the capillaries (tiny blood vessels) to shrink so less blood flows through the skin. This keeps heat from leaving your body.

As the wind whips around you, you begin to shiver. That's because a part of your brain called the *hypothalamus* is sending messages to your muscles telling them to increase their activity to create heat. Your muscles spring into action and make you shiver. The shivering helps create heat for your body.

You may also get goose bumps. Why? Your body is responding to the temperature by making tiny muscles pull the hairs on your body upright to trap warmer air next to your body. Unfortunately, humans no longer have enough body hair to do the trick.

If you get really cold, your body will direct the blood away from your hands and feet and toward your brain and other vital organs. This ensures that the most important parts of you stay warm the longest. That's why your hands and feet are often the coldest part of you. A finger or toe lost to frostbite is better than a cold heart, lungs, or brain.

Button Up!

When your body temperature drops below 95°F, a condition called *hypothermia* sets in. Hypothermia can cause confusion, speech difficulties, a loss of muscle coordination, and eventually death. So button up your coat and put on your hat and gloves. Better yet, stay inside when it's really cold.

It may be hard to imagine on that freezing walk to school, but soon summer will arrive bringing hot, steamy temperatures. Exercise or physical activity can also make your body temperature rise. So what does your body do to stay cool? Your brain will direct the capillaries near the skin to widen so heat is released into the air. That's why your face is red after exercise. The blood was rushing to the capillaries in your face to cool off.

Sweat is another cooling-off device. When sweat **evaporates** from the skin, it takes heat with it. The hotter a body gets, the more sweat it will produce in an attempt to cool down.

Try This!

Feel the cooling effect of evaporation. Sprinkle some water on the back of your hand. (Licking your hand will also work.) Sweat is mostly water, so water or saliva, which are also mostly water, will work as a substitute. Now blow on the moist area. (Your breath is acting like moving air, which speeds evaporation.) Can you feel how cool it gets? Try it again. This time blow on one wet hand and one dry hand, and compare the results. Can you feel the difference?

REGULATION IN ANIMALS
• • •

Location, Location, Location

Water animals have the luxury of living in surroundings where the temperatures don't change very much or very fast. Land-dwelling animals have to work harder to regulate their body temperatures since temperatures change more often and more quickly. Finding and moving to different locations can help these creatures control their temperatures.

Ever see a lizard sitting on a rock on a warm sunny day and wonder if it's trying to get a tan? Most likely it's regulating its body temperature. Sunny locations are hot spots for cold animals, especially **reptiles**. Garter snakes can raise their body temperatures as much as 50°F by sunbathing.

Snow can actually be a good **insulator** in cold weather. Mice, gophers, and shrews burrow in snow tunnels. Sled dogs in Alaska bury themselves in the snow to escape the chilling Arctic wind.

Since water is usually cooler than land, taking a dip is a good way to lower body temperature. Tigers have been known to beat the heat by soaking themselves in rivers. Elephants load their trunks with water and spray themselves.

On hot days, lizards and snakes will scurry under rocks to keep cool. Seeking shade or cover in the heat is a good way to escape scorching temperatures. Besides rocks, the ground beneath shady trees is often a popular place for overheated animals.

Some **amphibians** and reptiles bury themselves in sand or mud to escape temperatures that are too hot or too cold. They breathe through their skins while buried. Pigs are also known to roll around in mud to cool off.

When all else fails, some animals move to an entirely new location during harsh weather. This is known as **migration**. Monarch butterflies migrate to Mexico in search of warmer temperatures. Different species of birds and fish migrate to escape cold temperatures.

A Lot of Fluttering

The longest recorded monarch butterfly flight was over 1800 miles. This one-way trip began in Canada and ended in Mexico. That's a lot of fluttering!

Taking a Nap

Hibernation and **estivation** are two ways that animals protect themselves against cold or heat. Both processes result in a slowing of body functions to save energy and avoid extreme temperatures. The only difference is that hibernating is done in the winter and estivation takes place in the summer. Animals usually hibernate and estivate in nests, burrows, tunnels, dens, or other shelters that **insulate** them from the weather. Some animals both hibernate and estivate. Others do one or the other.

Warm-blooded animals that hibernate include bears, bats, chipmunks, raccoons, skunks, and groundhogs. Birds, such as poor-wills and nighthawks, also hibernate. Bees, worms, frogs, toads, lizards, turtles, snails, and snakes are cold-blooded hibernators.

Many reptiles, amphibians, and insects estivate. Lizards, turtles, snakes, snails, frogs, toads, bees, and worms are common estivators.

Nighttime Action

Nocturnal animals escape from the heat by doing most of their activities at night when it's cooler. They spend their days sleeping. Bobcats, tigers, leopards, foxes, and kangaroos are nocturnal land animals. Hippos are nocturnal animals that spend their days in the cool water and come out at night to find food. Many desert creatures are also nocturnal.

Shivering, Sweating, and Panting

Some animals have body functions similar to humans that help them stay warm or cool off. Most warm-blooded mammals can shiver. A few cold-blooded animals like bees and dragonflies can also shiver.

Many mammals have sweat glands in some parts of their bodies. Horses are the only other animals besides humans that have sweat glands all over their bodies. Camels' bodies are so well adapted to heat that they don't start to sweat until it's hot enough to make a human pass out.

Sweating Blood?

It was once believed that hippopotamuses sweat blood. It is now known, however, that these animals actually secrete a reddish oil when they sweat. This oil keeps their bodies moist in dry environments.

The way an animal breathes affects heat loss. Breathing through the mouth lets more heat escape than breathing through the nose. That's why panting increases the evaporation of heat. Dogs are famous for their panting. Other animals, such as coyotes, mountain lions, and birds, pant as well.

Hey, Those Are Cool Ears!

The large ears of elephants help keep the animals cool. Blood circulating through the large surface area releases heat into the air, cooling the elephant down.

Cover Up!

Many animals have special body parts to keep them warm. Birds have feathers that they can fluff up to trap warm air next to their skin. When birds get overheated, their lungs pump cool air through hollow spaces in their bodies called *air sacs*.

Other animals have hair or fur to keep them warm. Sea lions, polar bears, yaks, mountain lions, and other animals that live in cold environments are known for their thick coats. The arctic fox's fur protects it so well that the animal doesn't shiver unless the temperature drops to 90 degrees below zero.

Most mammals have a layer of fat under their skin to insulate them. A few have thick layers of fat, or blubber, that protect them from extreme cold. Whales, seals, walruses, and penguins are among these well-protected animals.

Huddle Together

What do snakes and puppies have in common? They huddle together to keep warm. Huddling, cuddling, and swarming are behaviors used by some animals to maintain body temperature. Puppies pile up on one another for warmth. Snakes huddle together underground. Cows and horses cluster together in cold temperatures. Honeybees swarm in large numbers and flap their wings to produce extra heat.

Taking Turns

Emperor penguins huddle in large groups to survive freezing temperatures. They rotate so that everyone has a turn to stand on the inside of the huddle where it's warmest.

Regulation in Plants

If you've ever left plants outside in blazing summer temperatures for too long or forgotten to bring plants in when temperatures fell below freezing, then you know how deadly temperature extremes can be for plants. Plants are perhaps the easiest victims of climate conditions. Unlike animals, they can't move to a shady spot when it's too sunny or burrow underground to keep warm. Instead they are left out in the open to deal with rising or falling temperatures.

So what can plants do to help regulate their internal temperatures? They can control their rate of transpiration or enter a state of **dormancy**.

Transpiration is the evaporation of water from plants. Most of this water is lost through the leaves during **photosynthesis**. Transpiration may have the same cooling effect as the evaporation of sweat in humans. Sunlight and warm temperatures increase the rate of transpiration. In hot, sunny areas, plants may lose more water than they can take in through their roots. This will cause the plant to wilt and possibly die. Slowing down the rate of transpiration enables a plant to survive until temperatures drop or it can take in more water. Many desert plants do part of their photosynthesis process at night or in the early morning to reduce the amount of water loss.

Dormancy also helps many plants survive winters. Dormancy in plants is similar to hibernation in animals. The plant slows down or stops its growth until conditions improve. Then the plant becomes active again.

Sleeping the Summer Away

Like to spend your summer vacation sleeping in? So do some plants. A few plant species, such as the drosera of western Australia, spend their summers in a dormant state. These plants can't adjust to warmer temperatures, so they rest during the hot months and wait for cooler weather to return.

chapter three

Behave Yourself!

HOW MANY TIMES HAVE YOUR PARENTS TOLD you to "behave yourself"? What did they mean? They meant you should act or respond in an appropriate way. The same is true for plants and animals. They must behave in ways that enable them to survive in their environment.

Unlike regulation, which occurs internally, behavior is the activity of an organism that can be observed from the outside. It involves movement or purposeful nonmovement, as when an animal stands perfectly still to avoid being spotted by a predator.

Behavior is a response to a **stimulus**. A stimulus is any factor that causes a reaction. It can be a change in the body or the environment. Smells, sounds, light, darkness, and temperature changes are all environmental stimuli that organisms react to. The action that an organism takes in response to a stimulus is its behavior.

How Many?

Stimuli is the plural form of *stimulus*. Sound is one stimulus. Odor and cold air are two stimuli.

I WAS BORN THAT WAY
● ● ●

Many behaviors are inborn. These are behaviors that an organism doesn't choose. They occur automatically without thought from the day an organism is born (or begins to grow). Inborn behaviors are passed on from parents to offspring, generation after generation.

The simplest inborn behaviors are reflexes. **Reflexes** are behaviors that happen automatically, like sneezing when pepper gets up your nose or dropping a hot potato. Blinking is a reflex. Reacting when startled is a reflex. Humans jump. Cats arch their backs. Octopuses squirt ink. Birds scatter or fly off.

An **instinct** is an inborn behavior present in a species that helps it survive and reproduce. Organisms are born with instincts. They don't learn them from their parents or other organisms.

When a spider spins a web, it is using instinct. Spiders don't have to be taught how to spin webs. They are born knowing how. Birds inherit the ability to make nests. Weaverbirds don't have to be taught how to weave 300 blades of grass into a basket-shaped nest. They do it instinctively. Termites build mounds because it is an instinct. Frogs lay their eggs in water because instinct tells them to do so.

Termite mound

LEARNING FROM EXPERIENCE

Not all behaviors, however, are inborn. Many animal behaviors are based on experience. These are called *learned behaviors*. When a parent teaches its young to hunt, fly, swim, or clean themselves, the offspring are practicing learned behaviors. When a pet learns to use a litter box, it's a learned behavior. Dogs and dolphins who perform tricks have mastered learned behaviors.

Most animal behavior is a combination of inborn and learned behaviors. For example, humans regulate body temperature through shivering, which is inborn, but they have also learned that putting on more clothes, rubbing their hands together, or lighting a fire will provide warmth as well. Instinct provides basic skills for survival, but learning what works best in a changing environment will make an organism most successful.

chapter four

I'm Hungry

THROUGHOUT HISTORY, HUMANS HAVE BEEN grouped according to the ways in which they get food. There are hunters, gatherers, fishermen, and farmers. Humans have learned to make tools, such as spears, fishing poles, rifles, and tractors, to help them get or produce food. Today, humans have become very successful at finding, growing, and making food. This is one reason the species has survived so well over time.

But plants and other animals don't have it as easy. When food runs out, they can't run to the grocery store or pull out their fishing gear. They can't plant their own fruits and vegetables or water themselves when the rain doesn't come. Luckily, an amazing variety of behaviors helps animals and plants get the food they need to survive.

ANIMALS

• • •

A Whole Bag of Tricks

Since animals can't make their own food, they need behaviors to help them get it from other sources. Being strong, fast, sneaky, and tricky are often helpful behaviors when chasing **prey**. A starfish proves its strength by wrapping itself around an oyster and prying the shell open. The starfish then pushes its stomach out of its own body and into the shell where it digests the oyster. Large wildcats like cheetahs, bobcats, and jaguars sprint toward prey, making the chase seem easy. Foxes are known for their cunning behavior. These quick animals dart about, often zigzagging, as they chase prey or outrun predators. The tricky snapping turtle lures fish with a pink tongue that looks like a tasty worm.

Determined to Have Termites for Dinner

Chimpanzees outwit termites by using tools. Chimps poke sticks into termite nests. When termites bite on the sticks, the clever chimps pull the sticks out and feast on the termites.

Poisonous behaviors are always good to have in a pinch. Pit vipers are snakes with pits, or holes, in their faces that sense the body warmth of prey. The snake sneaks up on its prey and injects poison into it. The poison doesn't kill the victim right away, so the snake follows the dying animal by smelling it with its tongue. When the animal finally dies, the snake enjoys a tasty meal.

Spiders don't have teeth to chew, so they use fangs to inject poison into an insect's body. Then they use special juices to turn the insect's insides into a liquid they can slurp out.

Flashy Tricks

Flashlight fish have organs under each of their eyes that contain millions of glowing bacteria. The fish can use these "lights" to search for food in dark ocean waters. One flashlight fish could light up a small room.

Where Are You?

Bats and dolphins use **echolocation** to find food. These animals send out high-pitched sound waves that bounce off objects in their paths. The length of time it takes for the echo to return pinpoints an object's location. Bats use echolocation to find moths and other insects, while dolphins use it to home in on fish.

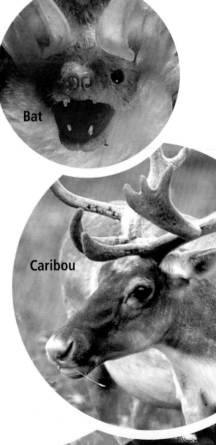

Bat

Caribou

Taking a Trip

Migration can help animals bring home dinner. Caribou and elk, for instance, migrate to warmer climates to support their all-plant diet. Many birds migrate to warmer locations when plants and seeds are buried in snow. Canada geese are highly visible migrators each fall. Small gnawing animals called *lemmings* often have population booms that wipe out the food sources in an area. When this happens, the lemmings go on mass migration in search of more food. Some whales migrate to colder oceans in the summer because the **krill** they like to eat is there.

Go Team!

Some animals use teamwork to get food. Social groups of animals sometimes share the jobs of hunting or finding food. Wolves hunt in packs. Lions live in groups called *prides*. The females of a pride usually hunt together. Wild chimps live in groups, or communities, of up to 100. They often work together to hunt colobus monkeys. When they get plenty of food, male chimps bang on tree trunks to call other chimps to share in the feast.

Some insects also work together in social groups. Bees send scouts to find nectar. Successful scouts return to the hive and do a dance to tell the other bees where to find the nectar.

South America's leaf-cutter ants grow their own food. Worker ants cut pieces of leaves from forest trees. They carry them back to the nest, where workers cut them into small pieces and spread them out. When the leaves become moldy, the ants harvest the mold and eat it.

Africa's army ants hunt in large swarms. They march through the forest in a long column like soldiers. Smaller worker ants carry food back to camp. Larger soldier ants march beside the column and defend it by swarming over any possible predators and tearing them to pieces with their powerful jaws. When army ants have devoured all the food in one place, they move to another.

Sometimes two completely different species help each other. This is known as **mutualism**. Sometimes both species get food. Sometimes only one gets food but helps the other in another way. Honey guide birds like to eat bee **larvae** and beeswax, but they can't open a beehive. So these birds lead honey badgers to hives. The badgers break open the hive and eat the honey, and the birds eat the leftovers. Another bird, the Egyptian plover, walks right into the open mouths of crocodiles to eat blood-sucking leeches from between their teeth. This helps the crocodile by removing the leeches.

Lampreys

Get Off Me!

A **parasite** is an organism that lives off and harms another organism. Fleas and lice are parasites that get their food from the bodies of other animals. Female mosquitoes need protein from blood for their eggs, so they bite humans and other mammals to suck their blood.

One of the creepiest blood-sucking parasites is the lamprey, a jawless fish. A lamprey has a round mouth filled with curved teeth. It uses these teeth to cut holes in other fish. Then the lamprey latches on with its mouth and sucks the blood and other body liquids out of the holes.

The tick should get the award for being the most patient parasite. A tick can sit and wait for up to three years for a mammal to walk by. Approaching mammals put out a smell that signals the tick to jump on it. The tick then gorges on the mammal's blood until it looks like a tiny balloon that's about to pop. Then the tick drops off, full and satisfied. Unfortunately, the ticks can cause irritation and disease in the mammals they dine on.

PLANTS

While you may not think of plants as moving and behaving, plants *do* respond to stimuli in ways that help them make or get food. Most plants make their own food through photosynthesis. This process requires sunlight, so most plants grow upward toward the Sun. Some turn toward the Sun as it moves in the sky to capture as much light as possible. Vines and other climbing plants wind their way up trees, other plants, or structures like trellises in order to get more sunlight. Morning glories, ivies, and clematis are some common vines. Rain forest climbing plants such as lianas and strangler fig trees must wind their way up through the layers of the forest to reach sunlight.

Morning glories

Water is also important to a plant's diet. Many plants in drier climates send roots deep into the ground to absorb water. Other plants shoot out thin roots near the surface to soak up as much moisture as possible. In order to conserve water lost during photosynthesis, a few plants can carry out parts of photosynthesis at night or during cooler parts of the day. Cactuses, pineapples, and some types of orchids can do this.

Try This!

Indoor plants may turn their leaves toward the window to collect as much sunlight as possible. See this for yourself. Gather a few houseplants. Put them in front of a window that receives direct sunlight. Watch the plants for a week. Do any of them have leaves that turn or bend toward the light? If so, rotate those plants a half turn and observe for another week. What happens?

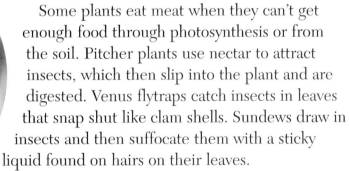

Sundew

Some plants eat meat when they can't get enough food through photosynthesis or from the soil. Pitcher plants use nectar to attract insects, which then slip into the plant and are digested. Venus flytraps catch insects in leaves that snap shut like clam shells. Sundews draw in insects and then suffocate them with a sticky liquid found on hairs on their leaves.

Plants can be parasitic too. Mistletoe is a plant that lives in tree branches and steals the tree's minerals, food, and water. Dodders are parasitic weeds that attach themselves to other plants and spread quickly. Rafflesias are giant plants that live off other plants on the rain forest floor.

chapter five

On the Defense

Humans have mastered the art of defense. They've developed physical practices such as the martial arts to conquer predators. They've invented weapons to fend off attackers. They have barriers, shelters, locks, and other devices to keep danger away. But animals can't install burglar alarms in their dens or nests. Plants can't fight off hungry insects with boxing gloves. These organisms, do, however, have their own forms of defense to protect themselves from harm.

ANIMALS

● ● ●

Safety in Numbers

Many species live in social groups that protect one another. Antelope and buffalo are safer in a herd than on their own. Prairie dogs live together in "towns" and even have special "watchdogs" that keep an eye out for trouble. Migrating geese fly in groups called *flocks*. Individual birds are less likely to get lost or preyed on this way. If a white-tailed deer is frightened, it flicks up its tail. This serves as a warning flag to other deer in the herd and everyone sprints to safety. Many fish swim together in groups called *schools*. This helps confuse predators who can't single out individual fish to attack. Some predators may even be fooled into thinking that the school is one giant animal that's too big to eat.

Catch Me If You Can

Speed helps many land animals escape predators. Red foxes avoid capture by wildcats and bears by running at speeds of up to 35 mph. The foxes are also good swimmers. Bush babies dodge eagles and snakes by leaping to safety. Mice are preyed on by many animals, so they are often seen scurrying, darting, or running at quick speeds. Flying fish escape harm by jumping out of the sea and gliding through the air with their fins stretched out like wings. They can glide for over half a mile until they reach safety. Even lizards can move amazingly fast when fleeing from danger.

If You Can't Outrun Them, Outwit Them

Some lizards shed all or part of their tails when attacked from behind. The tail keeps wiggling and distracting the predator while the lizard runs for safety. Later, the lizard grows a new tail.

If caught, kicking and butting are effective defensive actions. Horns, hooves, and antlers can be deadly when used with force. Zebra and deer have powerful kicks. Moose and sheep ram attackers with their antlers and horns. And no one wants to get in the way of an angry bull!

It's a Bird. It's a Plane. It's Superantelope!

Antelope in Africa can jump as far as 23 feet in a single bound when escaping predators.

Hide and Seek

Another behavior that is often used for defense is hiding. Animals hide under, inside of, and behind objects and other animals so they won't be spotted. Dashing behind trees, ducking in logs, crawling under rocks, and creeping through tall grass are all defensive actions.

Animals can also sometimes hide in plain sight just by holding perfectly still. Have you ever been so scared that you couldn't move? Many animals, such as deer, squirrels, and rabbits, freeze when frightened. It's a terrible survival tactic on the highway. But off the road, it is often a successful defensive behavior since many predators won't notice prey unless it's moving.

Have you ever heard of someone "playing possum"? Opossums and hog-nosed snakes pretend to be dead if they think they're in danger. This works because many predators won't eat something that's already dead.

Walking sticks are insects that vary their movement depending on the situation. Walking sticks look like thin green or brown branches. When they hold still or move very slowly, predators mistake them for sticks and leave them alone. However, on a windy day, a walking stick would stand out if it were the only thing not moving on a tree. So the insect moves its legs and quivers to appear as if it's shaking in the breeze.

Can I Borrow Your Shell?

Hermit crabs don't have hard shells to protect them like other crabs do. So these crabs borrow a shell left behind by a dead snail and hide inside it. As the crab grows, it moves into larger and larger shells.

Hiding Like a Vampire

Vampire squid hide from predators by pulling their webbed tentacles over their heads like a vampire pulling his cape over his face. The squid's spiky tentacles cover their bodies, making a shield that stops other animals from eating them.

Pufferfish

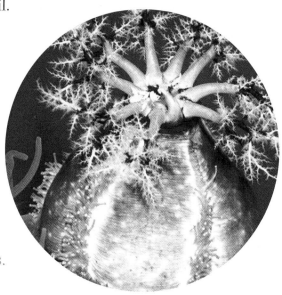

Australian frilled lizard

Stay Away!

Dogs bark. Cats hiss. Lions roar. Elephants trumpet. Rattlesnakes shake their tails. What's all this noise about? Sometimes animals try to scare predators with loud sounds or other behaviors that make them appear dangerous.

Many animals try to make themselves look bigger or scarier when threatened. Cats arch their backs. The hair on a dog's back stands up. Young owls will spread their feathers out to look bigger. Hawk moth caterpillars wave their bodies like a snake to scare away birds.

Pufferfish gulp water and blow themselves up like spiky water balloons to scare off attackers. If that doesn't work, at least the pufferfish is harder to eat.

The Australian frilled lizard doesn't usually look scary, but when it's alarmed, it spreads out a ruffle of skin around its neck and opens it up like an umbrella. This makes the lizard look big and scary. Then it hisses and thrashes it tail. Even dingoes, Australia's wild dogs, have been known to run from frilled lizards.

That's Disgusting!

If you're a predator with a queasy stomach, then you'd better stay away from the sea cucumber. When a sea cucumber senses danger, it squirts its insides out to distract the predator. Then it regrows the lost parts.

The Buddy System

Other species protect themselves through mutualism (working together with another species). For example, zebras, wildebeests, and gazelles often graze together for protection. Their large numbers ward off predators, and since they eat different types of grasses, they don't compete with one another for food.

Gazelles

Goby

Gazelles and ostriches often eat near one another. Their sight abilities are different, so one notices dangers that the other might miss. They both watch for predators and alert one another if an attacker creeps up.

A type of antelope called *impalas* team up with baboons. Herds of impalas act as watchdogs for troops of baboons. The fierce baboons provide protection for the impalas when predators approach.

Goby fish live with a nearly-blind species of shrimp. The two animals help one another. The shrimp build safe burrows for both animals to live in. When they go outside, the shrimp keeps a feeler on the goby's tail. When the goby senses danger, it swishes its tail and both animals hustle back into the safety of the burrow.

Good Buddies

Goby fish are also buddies with grouper fish. Gobies eat the parasites off grouper fish, which cleans the grouper. When predators approach, the grouper will let the goby hide inside its mouth until the danger passes.

Clownfish

Hermit crabs sometimes give sea anemones a ride on their backs. The anemone's stinging tentacles protect the crab from predators. The crab pays the anemone back by letting it eat leftover food scraps.

Suckerfish attach themselves to the backs of sharks for protection. While riding, the suckerfish eat shrimplike parasites on the shark, which helps the shark.

Clownfish live among the tentacles of sea anemones. Anemones have stinging tentacles that kill fish that swim too close. Clownfish, however, have special slime coverings that protect them from the poison. So the clownfish is able to live among the anemones, which kill any predators that might come after fish. In return, the clownfish help keep the anemones' tentacles clean.

Thanks for the Help

Sometimes a defense arrangement between two species only helps one of the species. The other species is neither helped nor harmed. This is known as **commensalism**.

Barnacles get a free ride from humpback whales. The barnacles attach themselves to the huge ocean animals for protection. They also find food while cruising through the water. The whales are unharmed but don't benefit from the arrangement either.

Pearlfish hide from predators by living inside sea cucumbers. They move in and out through the sea cucumber's breathing hole. The ungrateful pearlfish sometimes starts to eat the sea cucumber it lives in. When this happens, it's no longer commensalism. Now the pearlfish is a parasite.

Just Between You and Me

A relationship between two species is called *symbiosis*. Mutualism, commensalism, and parasitism are types of symbiosis. Each one is defined by whether one or both of the species benefits or is harmed by the relationship.

PLANTS

Many plants protect themselves by producing substances that are harmful or distasteful to predators. Milkweeds, for example, have an unpleasant taste, so most animals and insects avoid them. Grazing sheep, however, have been poisoned when eating these plants.

Some trees produce poison in their leaves only when they're being munched on. The leaves of Indian millet, also called *sorghum*, contain a poison that only becomes active if an animal takes a bite of a leaf.

Since plants can't move like animals, their defensive responses are limited. A few plants, however, have leaves that fold up when disturbed. This usually causes predators to lose interest and move on. The mimosa, also called the *sensitive plant*, has tiny leaves arranged in rows. When they're touched, the leaves close up in seconds. The Venus flytrap is another plant whose leaves close up as a means of protecting itself (and getting food).

Mimosa

Milkweed seeds

chapter six

From Generation to Generation

REPRODUCTION IS THE PROCESS BY WHICH AN ORGANISM produces new organisms of the same type. There are two ways to reproduce—asexually and sexually.

Organisms regulate and behave in order to survive and reproduce. If a species doesn't reproduce, it will cease to exist. So organisms must have successful behaviors for reproduction and the protection of young offspring.

ASEXUAL REPRODUCTION
• • •

In asexual reproduction, a new organism is created from a single parent. There are several forms of asexual reproduction. Some organisms simply split in two to create offspring. One bacterium splits into two bacteria. One amoeba splits into two amoebae.

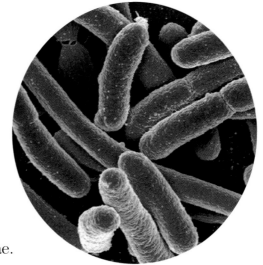

Other organisms have parts that break off and develop into new organisms. Yeast can form buds that break off and grow. Crabgrass and flatworms reproduce when part of the parent breaks away and grows. Gardeners can start a new geranium or African violet by planting a piece of the parent plant. Some plants, such as spider plants and strawberry plants, can reproduce by sending out special structures called *runners* that put down roots and become new plants.

Crabgrass

Flatworm

Even potatoes can reproduce asexually. A potato is a swollen stem called a *tuber*. It grows underground and stores food. If the eyes on a potato sprout, they can develop into new potato plants. You may see this if you keep potatoes in your kitchen cupboard too long.

SEXUAL REPRODUCTION

Most plants and animals reproduce by sexual reproduction. Sexual reproduction usually requires two parents. The male parent contributes a male reproductive cell called a *sperm*. The female parent contributes a female reproductive cell called an *egg*. When the two reproductive cells unite, a new organism develops.

Courtship

For sexual reproduction to take place in most species, males and females must find mates, or partners. It's much like dating in the human world. To do this, many animals have courtship behaviors. These are behaviors for attracting a mate.

Special dances or other physical actions are common courtship behaviors. Male spiders and seahorses dance for females they're interested in. The male peafowl (peacock) spreads his magnificent feathers into a huge fan and then shakes and makes loud rattling noises to impress female peafowl (peahens). The male bird of paradise attracts mates by dancing and showing off his beautiful feathers—sometimes while hanging upside-down in tree branches.

Male fiddler crabs wave their giant claws to attract females. Male pond turtles wave their long nails in front of a pretty female's face. Female wood ducks flick their bills over their shoulders. If a male is interested, he raises his wings and tail, turns, and swims away. Male leopards nuzzle females' necks to calm and attract them. American alligators share long periods of "cuddling" that includes touching snouts, bellowing, rubbing backs, blowing bubbles, and swimming together.

Male wood duck

Others species have mating calls or songs. Birds are famous for their beautiful mating tunes. Frogs croak to invite visitors of the opposite sex. Whales use high-pitched sounds to locate mates far away. Have you ever listened to crickets chirping outside your window on a summer's night? Those were male crickets trying to attract females.

Some species release chemicals called *pheromones* that a member of the opposite sex will respond to. Female moths have pheromones in their tails to lure male moths. Antelope secrete pheromones out of their eyes to find partners. Many animals spray urine to mark their mating territory.

Fighting is another way some species win a mate. Male elephants and gorillas will fight other males of their species for a female member. Wolves and dogs raise their fur, bare their teeth, and growl to tell other males to stay away from their chosen mate. Male blue tang fish impress females by using the bony spines at the end of their tails to duel.

A Dating Tip

One generous species brings gifts to impress their dates. A type of male nursery spider wins over females by bringing her a dead fly wrapped up in silk. (This is not recommended with human females!)

Great Beginnings

Successful courtship behaviors result in successful mating. It is then up to the new generation to survive. Some are helped by their parents' behaviors. Others are left to figure out the survival game by themselves.

Most fish, reptiles, and amphibians don't care for their young. The strategy is to lay lots of eggs and hope a few offspring survive to carry on the species. This is the main reproduction behavior of insects and other **invertebrates** too.

Baby sea turtles make their way to the ocean on their own. The mother turtle does not return once she buries the eggs in the sand.

The tilapia fish and the crocodile are two exceptions to this rule. The tilapia fish is called a *mouthbreeder* because when the female lays her eggs, she gathers them into her mouth to keep them safe. As soon as the babies hatch, she stands guard until they can take care of themselves. Mother crocodiles also stay with their eggs until they hatch. Then they carry the babies to water and watch over them for up to two years. Many male crocodiles remain with the family as well, protecting them from predators.

A Few Great Fathers

• A catfish father will keep eggs in his mouth until they hatch. This means he can't eat for weeks.

• Father seahorses carry eggs in a pouch. They keep them there for two months until the baby seahorses hatch and swim away.

• Emperor penguin dads warm and protect their eggs for 60 days. They keep the eggs on top of their feet under a feathered flap. The dads don't eat while they're standing guard. Many lose up to 25 pounds before the baby penguins hatch.

Egret family

Many birds and mammals knock themselves out looking after their offspring. Most birds stay with their eggs and keep them warm until they hatch. After the baby birds arrive, the parents care for and feed them until they can take care of themselves.

Many birds take great risks defending their young. For example, if an animal is threatening her young, the ringed plover mother bird fakes an injured wing. She staggers around looking helpless on the ground and lures the intruder away from the nest.

The cuckoo bird has found a way to get out of all that parental self-sacrifice while still making sure her offspring survive. By laying her eggs in another bird's nest, the cuckoo tricks the other bird into **incubating** and caring for her young.

Most mammals give birth to live young. Mammal babies are fed milk by their moms and are cared for by their parents until they are able to make it on their own. Most mammal parents teach their young how to care for themselves. Australia's marsupial mammals, such as kangaroos and wombats, even carry their young around in a pouch. The pouch protects the baby until it's fully developed and able to survive on its own.

Take a Deep Breath

When a baby dolphin is born, its mother quickly pushes it to the surface of the water and teaches it to breathe.

FROM GENERATION TO GENERATION
• • •

Behaviors for reproduction are vital to maintaining a species' population. If one generation can't pass on its behaviors to a new generation, the species will slowly (or sometimes quickly) disappear. The successful regulation and behaviors of one generation are meant to ensure many future generations of plants and animals on Earth.

INTERNET CONNECTIONS AND RELATED READING FOR REGULATION AND BEHAVIOR

● ● ●

Internet sites

http://yahooligans.yahoo.com/content/animals/
Want to know more about a particular animal and its behaviors? Then check out this Ranger Rick site that provides simple information on many mammals, fish, birds, insects, amphibians, and reptiles.

http://enchantedlearning.com/coloring/
Choose an animal by name, category, continent, or habitat and learn more about its unique behaviors.

http://kidsplanet.org/factsheets/map.html
Use these fact sheets on endangered animals to find out why their behaviors can't overcome predators or the environment.

http://www.nationalgeographic.com/kids/creature_feature/archive/
This National Geographic site for kids features many creatures and their behaviors and characteristics.

Books

Animal Dads by Sneed B. Collard III. This book describes how the males of different species help take care of their young. Houghton Mifflin, 1997. [RL 3.9 IL K–4] (3339901 PB 3339906 HB)

Animal Defenses: How Animals Protect Themselves by Etta Kaner. This book talks about what animals do to defend themselves when they are afraid. Kids Can Press, 1999. [RL 4.6 IL K–4] (3348701 PB 3348706 HB)

Habits of Desert Animals by Lynn M. Stone. Describes the behavior and daily lives of various desert animals. Rourke Book Company, Inc., 1997. [RL 4.7 IL 3–7] (5872906 HB)

Titles in the Animal Series:

Animal Relationships by Michel Barre. Gareth Stevens, 1998. [RL 5.5 IL 3–7] (5890706 HB)

Animal Senses by Michel Barre. Gareth Stevens, 1998. [RL 5.2 IL 3–7] (5890806 HB)

Animals and the Quest for Food by Michel Barre. Gareth Stevens, 1998. [RL 5 IL 3–7] (5890906 HB)

How Animals Move by Michel Barre. Gareth Stevens, 1998. [RL 5.9 IL 3–7] (5892006 HB)

How Animals Protect Themselves by Michel Barre. Gareth Stevens, 1998. [RL 5.8 IL 3–7] (5892106 HB)

•RL = Reading Level
•IL = Interest Level
Perfection Learning's catalog numbers are included for your ordering convenience. PB indicates paperback. HB indicates hardback.

Glossary

amphibian (am FIB ee en) cold-blooded animal that reproduces in water but lives its adult life on land (see separate entry for *reproduce*)

behavior (bee HAYV yer) the way in which an organism responds to its environment

commensalism (kuh MEN suh li zuhm) relationship between two species where one benefits and one is unaffected

dormancy (DOR muhn see) inactive state when growth and development of a plant slows or stops

echolocation (EK oh loh kay shuhn) locating an object using an echo (sound that bounces back to its source)

environment (en VEYE er muhnt) set of conditions found in a certain area; surroundings

estivation (es ti VAY shuhn) sleeplike state that animals go into during the summer to avoid extreme heat

evaporate (ee VAP or ayt) to change from a liquid to a gas

extinct (ex STINKT) having no members of a species alive

hibernation (heye ber NAY shuhn) sleeplike state that animals go into during the winter to avoid extreme cold

incubating (INK you bay ting) keeping a baby or egg warm until it's fully developed

instinct (IN stinkt) inborn behavior of a species

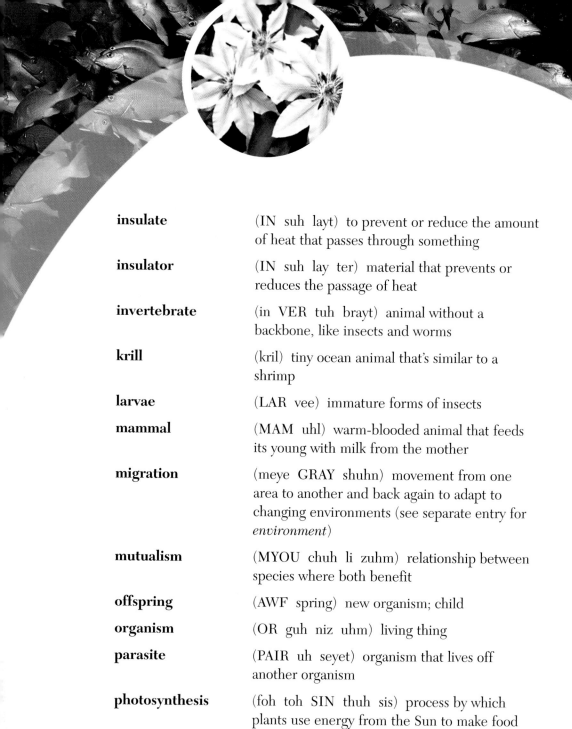

insulate (IN suh layt) to prevent or reduce the amount of heat that passes through something

insulator (IN suh lay ter) material that prevents or reduces the passage of heat

invertebrate (in VER tuh brayt) animal without a backbone, like insects and worms

krill (kril) tiny ocean animal that's similar to a shrimp

larvae (LAR vee) immature forms of insects

mammal (MAM uhl) warm-blooded animal that feeds its young with milk from the mother

migration (meye GRAY shuhn) movement from one area to another and back again to adapt to changing environments (see separate entry for *environment*)

mutualism (MYOU chuh li zuhm) relationship between species where both benefit

offspring (AWF spring) new organism; child

organism (OR guh niz uhm) living thing

parasite (PAIR uh seyet) organism that lives off another organism

photosynthesis (foh toh SIN thuh sis) process by which plants use energy from the Sun to make food

predator   (PRED uh ter) animal that hunts other animals for food

prey   (pray) animal that is hunted by other animals for food

reflex   (REE fleks) involuntary response to a stimulus (see separate entry for *stimulus*)

regulation   (reg yuh LAY shuhn) process of controlling the internal conditions of an organism

reproduce   (REE pruh doos) to make more organisms of the same species

reptile   (REP teyel) air-breathing animal that crawls or moves along the ground on its belly or short legs

species   (SPEE shees) group of living things that resemble one another and can reproduce to create more members of the group (see separate entry for *reproduce*)

stimulus   (STIM yuh luhs) factor that provokes a response in an organism

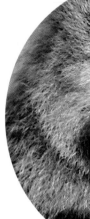

Index